1

De-Mystifying

the "Black Art" of

E^3

Byron D. Berman

Antenna on Cover:

Apologies to ETS-Lindgren, for the non-scale depiction of a very fine antenna, the 3106B, which I have personally used many times.

Library of Congress Cataloging-in-Publication Data is available.

Ellen Beth Berman
ISBN-13: 978-0692837146
ISBN-10: 0692837140

DEDICATION

To my father, of blessed memory, whose life was short, but dedicated to his family. He gave me a book called *The Boy Electrician* by Alfred P. Morgan (1940), which inspired me to want to understand electricity.

To my mother, of blessed memory, who believed in infusing everything with a bit of humor.

TABLE OF CONTENTS

BLACK ART?

"To me your field, E^3, is a black art."

--words of an outstanding leader of a major corporation

Black Art

An emotionally-based mystical practice often used to communicate with the mysterious realm of the unknown.

DOES IT TAKE A GENIUS?

Charles Proteus Steinmetz

(1865-1923)

A Prussian-born American mathematician, engineer, and professor, he worked for General Electric for thirty years and was called the "Wizard of Schenectady". Steinmetz is best known for his contribution in three major fields of Alternating Current (AC) Systems Theory: hysteresis, steady-state analysis, and transients. He was responsible for 200 patents, including the Induction Motor, Inductor Dynamo and Induction Furnace. Called the *Little Giant* because of his Dwarfism, Steinmetz was a true giant in the burgeoning field of electricity.

Excerpt from an article[1] about Charles Proteus Steinmetz, written by Gilbert King, August 16 2011.

"In 1894 he arrived in Schenectady, the place he would call home for the next thirty years, and his impact at General Electric was immediate. Using complex mathematical equations, Steinmetz developed ways to analyze values in alternating current circuits. His discoveries changed the way engineers thought about circuits and machines and made him the most recognized name in electricity for decades.

Before long, the greatest scientific minds of the time were traveling to Schenectady to meet with the prolific "little giant"; anecdotal tales of these meetings are still told in engineering classes today. One appeared on the letters page of *Life* magazine in 1965, after the magazine had printed a story on Steinmetz. Jack B. Scott wrote in to tell of his father's encounter with the Wizard of Schenectady at Henry Ford's River Rouge plant in Dearborn, Michigan.

Ford, whose electrical engineers couldn't solve some problems they were having with a gigantic generator, called Steinmetz in to the plant. Upon arriving, Steinmetz rejected all assistance and asked only for a notebook, pencil and cot. According to Scott, Steinmetz listened to the generator and scribbled computations on the notepad for two straight days and nights. On the second night, he asked for a ladder, climbed up the generator and made a chalk mark on its side. Then he told Ford's skeptical engineers to remove a plate at the mark and replace sixteen windings from the field coil. They did, and the generator performed to perfection.

---

[1] http://www.smithsonianmag.com/history/charles-proteus-steinmetz-the-wizard-of-schenectady-51912022/#lPk2lul2QkH82KD3.99

Henry Ford was thrilled until he got an invoice from General Electric in the amount of $10,000. Ford acknowledged Steinmetz's success but balked at the figure. He asked for an itemized bill.

Steinmetz, Scott wrote, responded personally to Ford's request with the following:

```
Making chalk mark
on generator:          $1.
Knowing where
to make mark:          $9,999.
```

Ford paid the bill."

Steinmetz and his contemporaries (Tesla, Einstein and others) at the Marconi wireless station in New Jersey.
Image courtesy of Wikicommons

How can someone who is *not* a genius, like Charles Steinmetz, hope to make important decisions concerning E^3?

How can a person in a leadership capacity, without even basic E^3 knowledge, make the correct program (contract) E^3 choices when needed?

BACKGROUND

The increasing awareness of Electromagnetic Environmental Effects' impact on electronics design has triggered the creation of an entire industry. After 41 years in the EEE/E^3 business, I am of the opinion that the challenges are more than just the physical phenomena, engineering documentation, and practices of EEE design & test. **There is also a hidden weakness inherent in the "company politic", the corporate culture**. This flaw inhibits the development of effective communication channels between the designated EEE Engineer, the management, and the government. **It has the potential to affect the contractual negotiations, whenever EEE standards are required**. The result is a breakdown in dialogue. This obstructs, or even makes impossible, the implementation of necessary system design attributes which successfully realize EEE compliance.

PURPOSE

De-Mystifying the Black Art of E^3 is designed to take the so-called, "Black Art", out of EEE and all of its subcategories (which includes EMI and EMC[2]). This is done without the force-feeding of a college curriculum that attempts to simplify the complexities of electro-physics. My book, instead, provides a way for the reader to get a general understanding or overview of all of it, without the technical education of an engineering student. Its purpose is to position anyone, knowledgeable or not in this field, to better understand EEE.

---

[2] EMI: Electromagnetic Interference. EMC: Electromagnetic Compatibility

1. THE QUESTION

What are Electromagnetic Environmental Effects?

My answer to this question is:

**Electromagnetic Environmental
Effects are phenomena
caused by the inadvertent
transfer of electromagnetic energy.**

YES. THAT'S IT!
THE WHOLE THING IN ONE SENTENCE!

The value of this seemingly simple answer lies in the fact that it is a first principle, a gateway to comprehending virtually all of the sub-disciplines that comprises EEE. It is a foundation for articulating scientific truth. Once that definition is fully understood, then the necessary control program for building an EEE-compliant product can be successfully developed.

Electromagnetic Environmental Effects (EEE) are described in a countless number of constantly-changing US and International specifications, standards, handbooks, textbooks, regulatory laws, certification programs, etc. That is the reason why the above definition is a powerful tool for quickly and easily gaining insight into a very large body of knowledge.

To gain a better appreciation for this definition, a review of another EEE definition is helpful. MIL-STD-464C, Section 3.4 defines Electromagnetic Environmental Effects[3,4] as follows.

---

[3] MIL-STD-464C, <u>Electromagnetic Environmental Effects Requirements for Systems</u>, 1 December 2010, is a US military standard.
[4] Some of the current acronyms for Electromagnetic Environmental Effects are EEE, E3, and E^3.

"The impact of the Electromagnetic Environment (EME) upon the operational capability of military forces equipment systems and platforms. E3 encompasses the electromagnetic effects addressed by the disciplines of:
Electromagnetic Compatibility (EMC)
Electromagnetic Interference (EMI)
Electromagnetic Vulnerability (EMV)
Electromagnetic Pulse (EMP)
Electronic Protection (EP)
Electrostatic Discharge (ESD) and
Hazards of Electromagnetic Radiation to Personnel (HERP)
Ordnance (HERO)
and volatile materials [fuels] (HERF).
E3 includes the electromagnetic effects generated by all EME contributors including Radio Frequency (RF) systems ultra-wideband devices High-Power Microwave (HPM) systems lightning precipitation static etc."

For people immersed in the business of Military EEE, this definition is useful, if not essential. However, for those individuals outside of the EEE community seeking some knowledge of EEE, such as Corporate Managers, Military Officers, as well as those engineers in need of greater clarification of EEE, this definition is neither practical nor insightful. Instead of fostering understanding, this definition can create a sense of greater frustration. Finally, this MIL-STD-464 definition addresses EEE only the United States (USA) Military, but does not serve the foreign military and global commercial markets. It lists a total of nine-plus undefined "disciplines," and depends upon a concept called Electromagnetic Environment (EME) which is also *undefined*, until a somewhat later section of MIL-STD-464C.

Let us now analyze this book's answer to "**The Question**", and then restate it in more descriptive terms.

2. ANALYSIS

> **Electromagnetic Environmental
> Effects are phenomena
> caused by the inadvertent
> transfer of electromagnetic energy.**

1. Phenomena: Observable occurrences.
2. Inadvertent: Unintended
3. Transfer: Move something from one location to another location.
4. Electromagnetic Energy: A class of energy associated with electricity.
5. Energy: The capacity to do work (move a force through a distance).

Rephrasing the answer:

> **Electromagnetic Environmental
> Effects are observable occurrences
> caused by the
> unintended movement of energy
> associated with electricity.**

Here are three examples. Each represents a different type of EEE:

EXAMPLE 1: Lightning

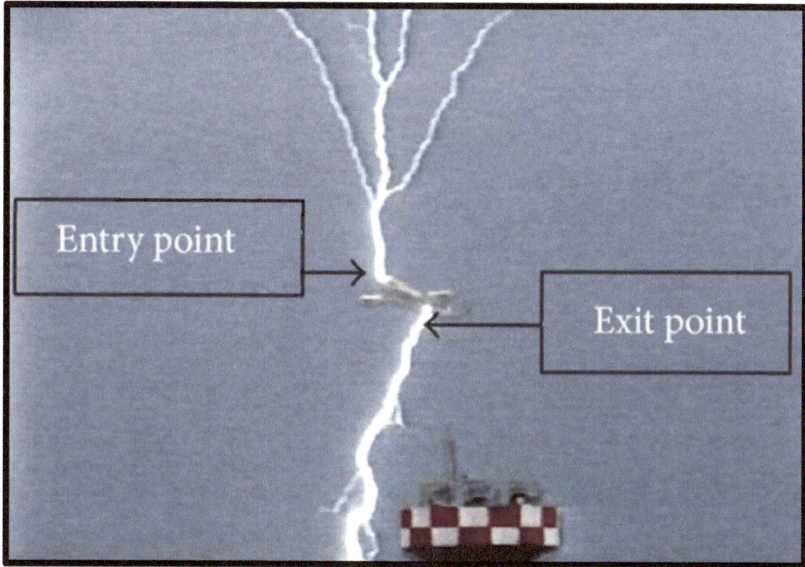

Video frame of a lightning strike to an aircraft on takeoff from the Kamatzu Air Force Base on the coast of the Sea of Japan during winter.
Courtesy, Z. I. Kawasaki[5].

1. Phenomenon: Lightning. The naturally occurring transfer of a large quantify of electrical charge from cloud-to-cloud, or cloud-to earth (and vice versa, depending on polarity).
2. Unintended: The strike occurs due to the presence of the aircraft, when it electrically reduces the resistance of the discharge path.
3. Transfer: Conduction [See entry and exit points shown in picture.]
4. Electromagnetic Energy: Lightning can transfer anywhere from one billion to 10 billion joules of energy (via electric charge). One joule per second is the equivalent of one watt of power.

---

[5] M. A. Uman and V. A. Rakov, "The Interaction Of Lightning With Airborne Vehicles," *Progress in Aerospace Sciences*, vol. 39, no. 1, pp. 61–81, 2003.

EXAMPLE 2: Electromagnetic Interference

View from the helipad of a recently completed tower of a
world-renowned teaching hospital

1. Phenomenon: Electromagnetic Interference: It was observed
    that the arrival of a helicopter on the roof correlated with the
    perturbation of EEG signals being measured in the Epilepsy
    Unit, immediately below the helipad.
2. Unintended: No one anticipated that the electromagnetic
    emissions associated with the helicopter would interfere with
    the EEG machines, preventing good readings.
3. Transfer: Radiation from the helicopter to the EEG machines.
4. Electromagnetic Energy: Electromagnetic radiation from the
    helicopter avionics coupled to the sensitive circuitry of the
    EEG, and induced an undesired electrical noise level. This
    was proven to be the fact, when RF shielding was retro-fitted
    into the Epilepsy Unit to correct the problem, a costly
    miscalculation.

EXAMPLE 3:   Electromagnetic Pulse

Nuclear Detonation

1. Phenomenon: Electromagnetic Pulse: Transient electromagnetic field radiated from a nuclear detonation.
2. Unintended: Initially, the electromagnetic radiation (non-ionizing) was not known. Only the explosive yield and associated damage was of interest and controlled at some basic level.
3. Transfer: Radiation from the area at, or above "ground zero" travels to a distant population center.
4. Electromagnetic Energy: An extremely large electromagnetic radiation pulse induces large electric transients in electrical and electronic systems, either damaging them or impairing their operation.

3. APPLICATION OF THE DEFINITION

We are ready to use the definition of EEE in Chapter 1 to define the MIL-STD-464 sub-disciplines:

1. **Electromagnetic Interference (EMI):** The inadvertent transfer of electromagnetic energy between electronic systems, characterized by the unacceptable performance of one or more of the receiving systems[6].

2. **Electromagnetic Compatibility (EMC):** A design goal. A state of being that exists between electronic systems, whereby the inadvertent transfer of electromagnetic energy between those systems is sufficiently controlled, as evidenced by the acceptable performance of all of the systems.

3. **Electromagnetic Pulse (EMP):** The inadvertent transfer of electromagnetic energy[7], sourced from a nuclear detonation, causing permanent damage, or unacceptable performance in electronic systems.

4. **Lightning Direct:** Damage phenomenon or unacceptable performance, caused by the inadvertent transfer of the lightning strike conducted electromagnetic energy to an electronic system.

5. **Lightning Indirect:** The inadvertent transfer of electromagnetic energy, induced from a lightning strike, to electronic systems, causing unacceptable performance or permanent damage.

---

[6] This also applies within a system, subsystem, assembly, module, or circuit.
[7] Non-ionizing.

6. **Hazards of Electromagnetic Radiation to Ordinance (HERO):** The inadvertent transfer of electromagnetic energy to weapons and other pyrotechnics at levels that cause detonation.

7. **Hazards of Electromagnetic Radiation to Personnel (HERP):** The inadvertent transfer of electromagnetic energy to people at levels that cause deleterious physiological effects.

8. **Hazards of Electromagnetic Radiation to Fuels (HERF):** The inadvertent transfer of electromagnetic energy to fuels at levels that cause ignition.

9. **TEMPEST:** The inadvertent transfer of unencrypted classified intelligence via the media of electromagnetic energy, mechanical energy[8], or acoustic energy at levels which permits said intelligence to become available to unauthorized personnel.

10. **EMCON: Emissions Control:** The control of the intentional and inadvertent transfer of electromagnetic energy from military platforms to levels sufficiently small, facilitating obscuring the presence of that platform from detection.

11. **Electrostatic Discharge (ESD):** The inadvertent transfer of electrical charge to electronic systems, causing permanent damage or degraded performance.

---

[8] Slightly broader definition.

4. ELECTROMAGNETIC ENVIRONMENT

The words *transfer between* suggests that there exists a physical region, or location between transfer points. When injected with electromagnetic energy, that region or a composite of such regions, form what is often referred to as the **Electromagnetic Environment (EME)**.

A classic example of EME is the region on and around a naval ship, such as the (now decommissioned) CGN 38, the USS Virginia. Partial contributions to that environment were the AN/SPS-48, 3-D Air search radar, the AN/SPS-40, 2-D Air search radar, the AN/SPS-55 surface search radar, the AN/SPQ-9 gun fire control radar and the AN/SPG-51 Missile fire control radar. These subsystems emit different kinds of electromagnetic energy. During the operation of these subsystems, electronics and people, if erroneously located in identified "keep-out zones" of that ship, could be subject to malfunction and physiological damage, respectively.

USS Virginia (CGN-38)

5. IMPACT

You may be: writing an RFQ[9], evaluating bids, writing a proposal, managing a prime contract, or managing a subcontractor for a military project. Sub-disciplines of EEE will ultimately be levied as requirements.

Nineteen sixty-eight was the year that MIL-STD-461A was published. Its topic was Electromagnetic Interference (EMI). There had been other standards published prior, but MIL-STD-461, currently going into revision G 48 years later, is predominant in the US Military procurement process. True, there are many other requirements of related topics that are used, including the umbrella-document MIL-STD-464C which focuses on Electromagnetic Environmental Effects (EEE). However, the longevity of MIL-STD-461 is testimony, in itself, as to the fundamental need for Electromagnetic Compatibility (EMC), affected by EMI compliance.

If the leadership handling a military contract has no knowledge of EEE, even as much as seeing a need for hiring an independent consultant, then there is a 99% probability that EEE will create major problems for that contract. Major, in this case, means financial, schedule or interference problems, just to name a few. A corporate reputation, plagued with over-runs and non-compliance, takes years to overcome.

---

[9] Request for Quote

Figure 1 shows a conceptual plot of the percent completion (solid line) of a military contract and typical milestones in a contract life-cycle. The probability of implementation of EEE design attributes (dashed line) decreases as the product matures. This notional sketch assumes that everything needed to be on the contract was done.

If, for some unknown reason, EEE requirements are missing, and their absence is identified (for example) during the Preliminary Design Review (PDR), for example (Notional Figure 2), then the probability of implementation is greatly reduced. Most likely, the later in time that this realization occurs, the greater will be the distortion in the completion curve in order to affect implementation.

Admittedly, this is not unique to EEE. Nevertheless, contract costs, product deliveries, and product performance during the mission are all military-industrial business issues that can be affected in a major way by the proper or improper handling of EEE requirements.

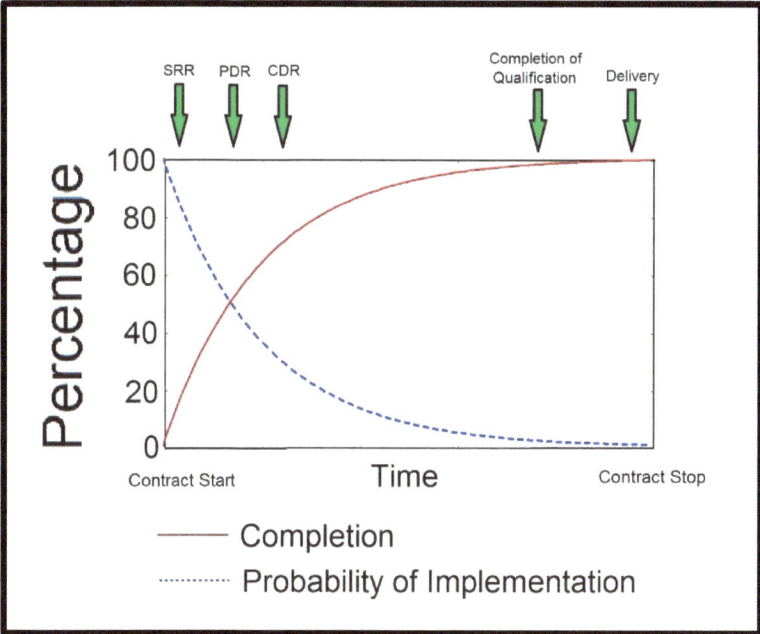

Figure 1 Effect of Time on the Implementation of EEE Design Attributes

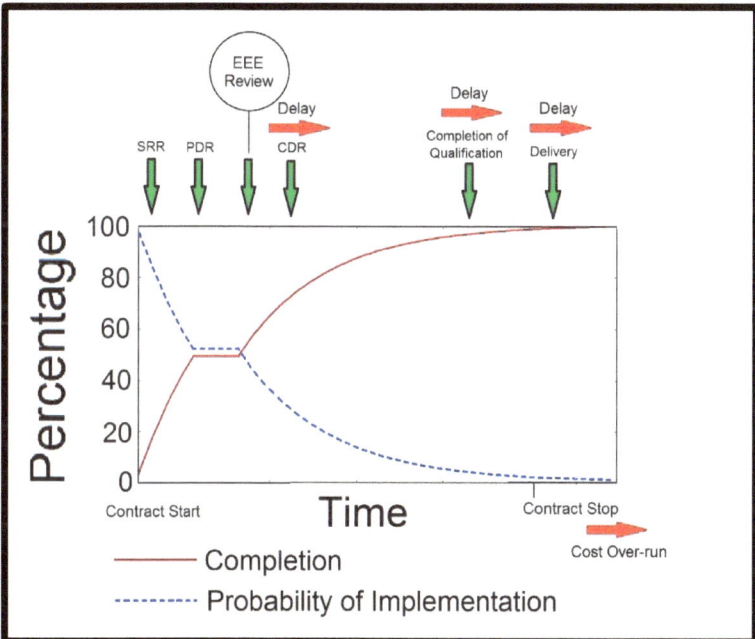

Figure 2 Life Cycle Distortion Due to Delay of EEE Design

6. UNDERSTANDING THE EEE "BUSINESS"

- **What is the fundamental core?**
My perception is as follows:

> **The Electromagnetic Environmental Effects "Business" is one of controlling the inadvertent transfer of electromagnetic energy.**

The core of the EEE business is found in this definition. It involves two ideas: the idea of **transfer,** and the idea of **control**. Here are some examples of how these terms apply to *other* businesses:

1. In the concrete business, one transfers the cement (wet or dry, with aggregate or not) from the factory of origin to either a user, or a distributor.

2. In the Girls Scouts, a parent picks the kids up from their homes, and drives them safely to their meeting at the community center in a timely manner.

3. A company, such as Federal Express, transfers items within the continental United States, from Point A to Point B **overnight**.

In each of the three civilian examples given above, the transfer is easily identified and is intentional. However, in the EEE Business, the focus is the control of transfer of **inadvertent electromagnetic energy** which permits only *specific types* and *quantities* of electromagnetic energy (if any at all) to reach the protected receive-points (or receptors) from the points of emanation (or sources). This idea is the **foundation of this business** that must be recognized, in order for EEE to be understood.

What is a classic example of an inadvertent transfer?

An electrical cable, when in use, not only conducts electricity from point-to-point, but also creates some form of electromagnetic field in its vicinity. The dynamics of the electricity being conducted affect the characteristics of the associated field. This field is all but irrelevant, except when it *inadvertently interacts with some other electronics* (or structure, or organism) in an unacceptable way.

Reciprocally, when this cable is immersed in an electromagnetic field created elsewhere, this other field is all but irrelevant except when it *inadvertently couples electromagnetic energy onto the cable* causing malfunction. **In either case, when the interaction of the electromagnetic field results in an unacceptable condition, the result is one of electromagnetic interference (EMI), and the transfer of electromagnetic energy is inadvertent.**

For military electronics, identification of inadvertent energy transfer is not easy. Ideally, objective evidence as to the ***precise nature*** of the inadvertent transfer(s) should be identified <u>prior</u> to implementing an effective method of controlling transfer. In reality, this can be difficult, if not impossible[10].

---

[10] For example, this could require a detailed investigation of combat mission forensic evidence.

• What makes the EEE business so very difficult?

The forms of electromagnetic energy may be *intentional by design, or inadvertent as a byproduct of a design's function.* The mathematical relationship between the physical size of the design to the wavelength of the energy identifies if the phenomenon is discrete or distributed.

Electromagnetic energy, distributed as a sine wave over physical space, exhibits a wavelength as shown below:

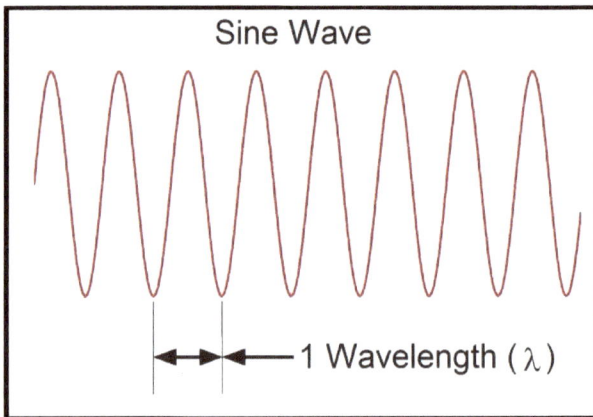

As a "rule of thumb", **when the physical size of the design is less than one-fourth of a wavelength of the energy, then the relationship is categorized as discrete. When the physical size is larger, the relationship is considered distributed.**

For example:

The wavelength of visible light is extraordinarily small, compared to the much larger physical size of the light bulb. Light is the distributed phenomenon.

Yet a light source, such as the light bulb, is a discrete device, because its physical size is small compared to the wavelength of the 60 Hz electricity.

Adding to the complexity of the relationship of the design to the electromagnetic energy, **mechanisms associated with the transfer of electro-magnetic energy are often a mix of both discrete and distributed physical phenomena**. This occurs when the physical sizes of various components of the design range from smaller to larger than one-fourth of a wavelength of the energy.

• How are discrete and distributed phenomena related to the idea of transfer?

There are many devices which enable electromagnetic energy transfer. One such device is called an antenna. In the case of the cell phone, a rather small antenna takes high frequency electricity from the transmit circuits.[11] It then transforms the electricity into electromagnetic radiation, which travels over long distances.

ETS-Lindgren
3106B Antenna

Antennae are used for **intentional transfer**. However, electromagnetic fields are to be found wherever there is electricity. This applies to both the discrete and distributed phenomena. **The object of this discussion is not necessarily radiation, but fields—electromagnetic force fields distributed throughout physical space.** These fields can interact with other circuits, and cause the transfer of electromagnetic energy to them, whether *they are supposed to or not.*

---

[11] The electrical power that is used in the home oscillates (changes harmonically) at 60 cycles per second (Hertz is the name of the units of frequency). The cell phone electricity oscillates about 15 million times faster.

When the source of the electromagnetic field is a circuit that is not designed to radiate, then the quantification of the circuit-to-field relationship is more difficult. It can be found in solutions to Maxwell's equations, but not necessarily in books or technical articles. **Analytical or numerical models may be required to represent the physical relationship.** (One such model is shown below.) This may include <u>deriving</u> new assumptions, approximations, and worst-case evaluations—all within the capability of the EEE Engineer.

Azimuth Electric Field Component of the Stratton (Dynamic) Magnetic Dipole Model

$$E_\phi = \frac{k^2}{4\pi} \cdot \sqrt{\frac{\mu}{\varepsilon}} \cdot \left[\frac{1}{R} + \left(\frac{j}{k \cdot R^2} \right) \right] \cdot A \cdot I \cdot \exp(j \cdot k \cdot R) \cdot \sin(\theta)$$

where: A = area
k = wave number
R = distance
I = current

$j = \sqrt{-1}$
θ = elevation angle
μ = permeability
ε = permitivity

- **What is the function of the EEE engineer?**

Kitja Kitja/330026990/Shutterstock

EEE requirements provide a **tailorable**, experiential data base of electromagnetic energy transfers, which can be applied to the military system and mission of interest. An EEE requirement is "tailored" when it is modified to reflect physical conditions associated with the functioning of the system, and its tactical environment. Once the transfers are identified, a method(s) of **transfer control** can then be determined.

In situations where the identification is difficult, EEE engineers and consultants are used to facilitate the identification process. They enable a truly comprehensive understanding of the EEE problem and provide direction for realizing the best results.

A good EEE engineer has:

1. A broad knowledge base of electrical engineering principles, being able to:
   a. grasp the design of the system.
   b. identify and quantify the interactions of unintended electromagnetic energy and the system.
   c. distinguish between the discrete and distributed phenomenon so that the appropriate transfer controls can be determined
2. Sufficient experience to select appropriate transfer control methods. (See following page.)
3. The capability to effectively communicate with other engineers, as well as with management. (See "Parting Thought" at end of book.)

IMPORTANT NOTE

Unless the EEE engineer, "coming in off the streets", has intimate knowledge of the complex system, or is given complete cooperation and access to program personnel and resources, he will not be able to provide support that will yield an EEE-compliant design. Such a person cannot assimilate hundreds, if not thousands of hours of engineering effort, in a matter of weeks. Management authorization is needed to streamline this process.

• What electromagnetic energy transfer controls are available?

Specialized products are manufactured and sold for the purpose of controlling the inadvertent transfer of electromagnetic energy. This is an important part of the EEE business. The EEE engineer's knowledge of the application of these products is crucial for implementing a successful method of control. Examples of such products include: EMI filters, EMI gaskets, specialized metal meshes, metalized paints and caulks, continuously-woven metal braids, lightning rods, spark gaps, and certain solid state circuit components.

If the product selection process were purely technical, it would be difficult. However, it often becomes a veritable "turkey shoot", a process with the inherent warning *caveat emptor*, let the buyer beware. The following issues[12] (among others) can present themselves during a selection process:

1. The product specs are incomplete.
2. The product specs are inaccurate.
3. The product specs or the supplier itself misrepresents the product.
4. The product specs are outdated, and the supplier builds to a set of fabrication instructions, without the knowledge of the engineering that went into the original design.
5. The supplier uses test methods which are flawed.
6. The product does not meet the targeted environmental conditions.

These kinds of issues may be avoided by the judgement of an experienced engineer. When the EEE engineer is very familiar with a system design, he may choose to use certain features of the design itself as a control.

---

[12] Not unique to EEE-related products.

Physical attributes of the design can, in a timely manner, be cheaply manipulated to control the transfer of electromagnetic energy. Filter-port placement, chassis-wall thickness, circuit ground-layer placement, differential circuit usage, shielded chassis vent or weep holes, shielded cables, and controlled impedance cables can contribute to this capability. This, however, requires intimate knowledge of the product being designed, and excellent communication between the EEE engineer and the product design team.

A very early quantitative assessment of the EEE risks guides the direction of system design changes. These can become prohibitively expensive as the design matures. Time, for this purpose, is measured by the milestones of the system life cycle.

For the purposes of this book, I am defining system life-cycle as the collection of chronological events that characterize the state of the system from concept, through fabrication, to delivery.

- **How do we know what quantity of control is really necessary?**

What is required may be identified in the EMIB.

> **An Electromagnetic Isolation Budget (EMIB) is a quantitative engineering tool for synthesizing an EEE-compliant design.**

Engineering analysis is used to derive an Electromagnetic Isolation Budget (EMIB)[13] from the quantification of the inadvertent transfers. When the EMIB is physically realized in the design, it diminishes "phenomena caused by the inadvertent transfer," to the point of **irrelevancy.**

More often than not, there may be several Electromagnetic Isolation Budgets associated with a single product:

1. A budget associated with conducted transfer phenomena,
   a. *From* the product.
   b. *Toward* the product.
2. A budget associated with radiated phenomena,
   a. *From* the product.
   b. *Toward* the product.
3. Budgets that vary with frequency or a number fixed in time.

Reconciliation of these different budgets may be needed, when only *one group* of transfer controls is feasible for all. Why this is true, is beyond the scope of this book. Test data may be called for.

---

[13] Back in the mid 1980's, as lead EMC engineer on a contract, I coined the phrase EMI Isolation Budget (EMIB), which is a quantitative tool for synthesizing an EMC design. The same holds true for the more general term Electromagnetic Isolation Budget (EMIB).

- **What is the philosophy behind the derivation of an Electromagnetic Isolation Budget?**

> **The EMIB is the ratio in decibels of the parameters of transferred electromagnetic energy to those same parameters of system-compatible energy levels.**

The most common parameters are power, voltage, current, power density and field strength. By applying methods of engineering physics, we derive the specific quantitative values of both the electromagnetic energy transferred and the acceptable energy.

In terms of product EEE design,

1. The more EEE control products that are incorporated into the design,
   a. the higher the product cost.
   b. the larger the product weight.
   c. the more complex the physical design.
2. The greater the accuracy of the EMIB,
   a. the more cost-effective is the EEE-compliant design.
   b. the better matched is the quantity of protection.
   c. the greater the potentiality of the system for use in different environments.

7. E3:

CONFUSION TO COMPLIANCE

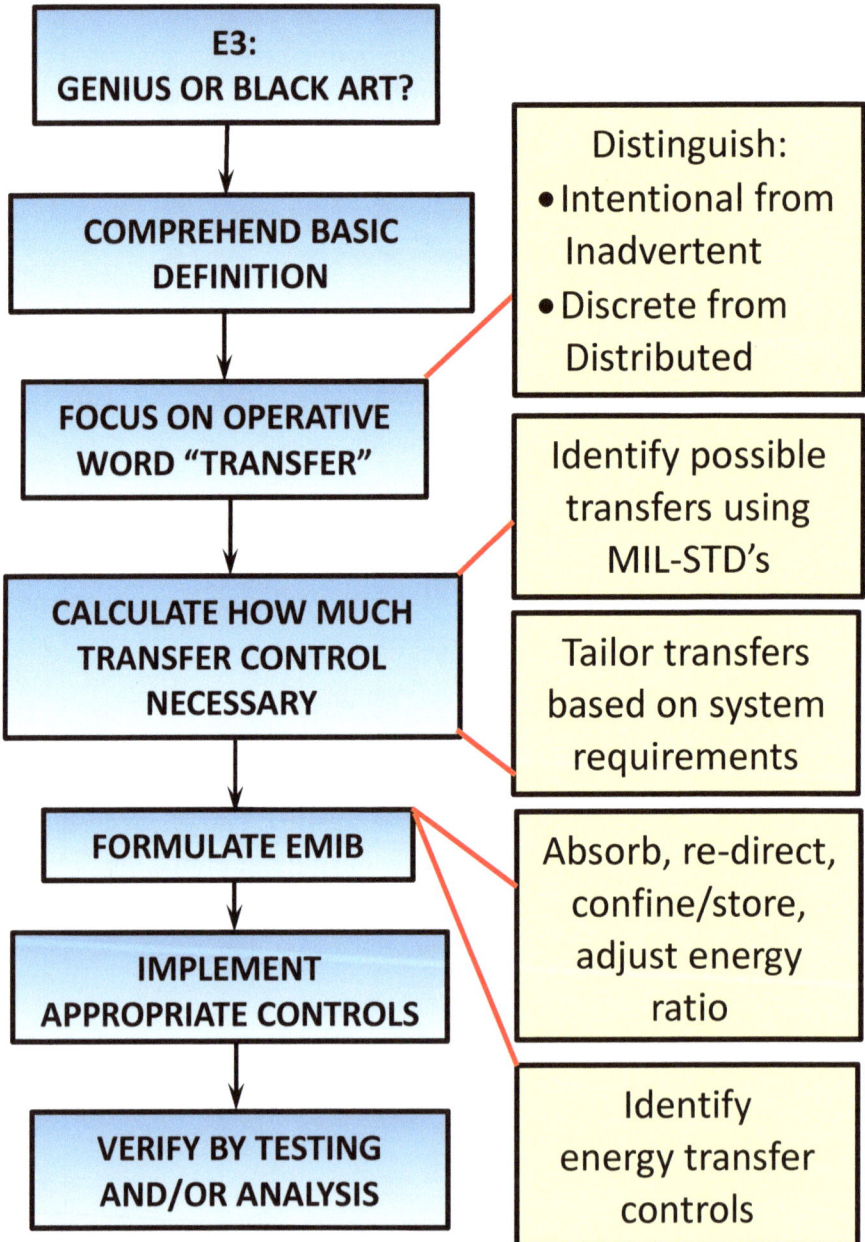

```
┌─────────────────────────┐
│         E3:             │
│ GENIUS OR BLACK ART?    │
└─────────────────────────┘
           │
           ▼
┌─────────────────────────┐        ┌──────────────────────────┐
│  COMPREHEND BASIC       │        │      Distinguish:        │
│     DEFINITION          │        │ • Intentional from       │
└─────────────────────────┘        │   Inadvertent            │
           │                       │ • Discrete from          │
           ▼                       │   Distributed            │
┌─────────────────────────┐        └──────────────────────────┘
│  FOCUS ON OPERATIVE     │
│   WORD "TRANSFER"       │        ┌──────────────────────────┐
└─────────────────────────┘        │   Identify possible      │
           │                       │   transfers using        │
           ▼                       │      MIL-STD's           │
┌─────────────────────────┐        └──────────────────────────┘
│  CALCULATE HOW MUCH     │
│  TRANSFER CONTROL       │        ┌──────────────────────────┐
│    NECESSARY            │        │   Tailor transfers       │
└─────────────────────────┘        │   based on system        │
           │                       │   requirements           │
           ▼                       └──────────────────────────┘
┌─────────────────────────┐
│   FORMULATE EMIB        │        ┌──────────────────────────┐
└─────────────────────────┘        │  Absorb, re-direct,      │
           │                       │  confine/store,          │
           ▼                       │  adjust energy           │
┌─────────────────────────┐        │  ratio                   │
│    IMPLEMENT            │        └──────────────────────────┘
│ APPROPRIATE CONTROLS    │
└─────────────────────────┘        ┌──────────────────────────┐
           │                       │     Identify             │
           ▼                       │  energy transfer         │
┌─────────────────────────┐        │    controls              │
│  VERIFY BY TESTING      │        └──────────────────────────┘
│  AND/OR ANALYSIS        │
└─────────────────────────┘
```

COMPLIANCE!

The transfer control(s) selected must constrain the induced energy to levels below those which cause degradation of system performance. When these controls are implemented, the EEE Design becomes compliant.

A PARTING THOUGHT

Confusion over the process of developing the Electromagnetic Isolation Budget (EMIB) contributes to the idea that EEE is a *"Black Art."*

Frank Morgan as the Wizard of Oz.
(MGM, 1939)

I hope this book has brought the engineering specialty of EEE out from behind the opaque curtain of mathematics and engineering jargon, "de-mystifying it", and bringing it into a better light.

With our youngest grandson, Davey

Byron Berman's career spans over 41 years of hands-on experience in Electromagnetic Compatibility (EMC) design, Electromagnetic Interference (EMI) analysis and EMI testing of complex military and commercial electronic systems. His range of expertise includes classical electro-physics, engineering mathematics, microwave circuits and circuit theory.

Mr. Berman works with high level proposal & contract documentation, as well as with the "nuts & bolts" of EMC design and testing. Even though Mr. Berman has received the coveted Eagle Award for "Blazing A Path To Mission Success", from the United States Government, he still considers the greatest compliment, that of an EMI technician, "It's always fun working with you, Byron!"

www.ingramcontent.com/pod-product-compliance
Lightning Source LLC
Chambersburg PA
CBHW041720200326
41521CB00001B/138